Cross Section of

DRAM

(Dynamic Random Access Memory)

Kung Linliu, Ph. D.

Author: Kung Linliu, Ph. D.

He got a Ph. D. degree in polymer physics and chemistry at Stony Brook, State University of New York (SUNY, now renamed as Stony Brook University, or SBU) and a bachelor's degree in chemistry from National Taiwan University.

He has also served as an Adjunct Professor in several National Taiwan Universities and taught the semiconductor IC process and MEMS courses for several years. He has written more than 30 academic and conference papers.

He has industry experience for more than 26 years, with more than 190 patents of USA, Taiwan (ROC) and China (PROC); he has founded two companies making inkjet printhead chip and ink cartridge, industrial inkjet printer, inkjet nano silver and copper ink, water and solvent based color (CMYK) inks, and special solvent inkjet inks for various substrates such as sapphire and aluminum nitride

(AlN for LED).

Before founding the first company, he has worked in the field of semiconductor DRAM, FLASH, MROM memory and logic ICs. Then he worked as a research fellow for MEMS-related research and development (R&D) and business start-up related technical transferring for printing and inkjet printhead at National Synchrotron Radiation Research Center (NSRRC).

His experience in academic research has about the topics of inkjet droplet forming simulation of piezoelectric inkjet printhead with micro-fluid dynamics, the study on the growth model of thermal bubbles under the influence of gravity and hydrophilic properties, and the delivery of drugs with MEMS nebulizer; the study on polymer phase separation and related critical phenomena by using synchrotron radiation small angle X-ray scattering (SAXS) and centrifugal force, and the viscosity of ultrahigh molecular weight polyethylene (UHMWPE), and the motion phenomena of solid pellets in liquid polymers under high centrifugal force.

He has published more than 85 books including more than 58 are in the professional fields of semiconductor, IC, inkjet printing, MEMS, printed electronics, RFID, FPCB, micro-LED and display and anti-fake hologram patterning related products, and so on.

He is now as a VP (vice president) of K-laser Technology, Inc.   And as a CTO (chief technology officer) of K-laser companies group (including worldwide sales channels and factories with two domestic subsidiary companies OPTIVISION TECHNOLOGY INC. and INSIGHT MEDICAL SOLUTIONS INC.) for anti-fake holographic patterning products and also the hologram papers and plastic films for packaging and packaging materials.

He is doing a two year research and development (R&D) project with 40 million $NTD and has budget from government (financial subsidy) for the hologram product related improvement of K-Laser Technology, Inc. This project is initiated from the cross-fields including technology of MEMS, inkjet printing technology and semiconductor IC process. Where chemistry, physics, materials, semiconductor IC and so on are incorporated in the research project for the advanced process as a pioneer investigation in this field.

Preface

A detailed cross section of DRAM is presented in this book.

DRAM is abbreviation of Dynamic Random Access Memory. DRAM is volatile memory used for electronic devices such as personal computer, cell phone, pad, etc. DRAM (Dynamic Random Access Memory) becomes an important component for electronic devices is after Bill Gates and Paul Allen found Microsoft Corporation which makes the personal computer software operation system DOS (disk operation system) in 1975.

Nowadays, the memory unit is improved and has a lot of progress from K (kilobyte or KB) to G (gigabyte or GB). The DRAM memory IC is a key component of personal computer and the price of it is lower than that of the CPU IC.

Last year, at the conference of 2020 International Electron Devices Meeting (IEDM), Imec presented a paper on a novel capacitor-less DRAM cell architecture, which is DRAM cell with 2T0C cell structure.

Kung Linliu
Hsin-Chu
Feb.
2021

Dedicated to ~

My parents, wife and children

Contents:

Chapter 1. DRAM IC structure

The chapter is to depict the structure of DRAM IC briefly.

1-1. What is DRAM?

DRAM is abbreviation of Dynamic Random Access Memory. DRAM is volatile memory electronic devices such as personal computer, cell phone, pad, etc. DRAM is used for loading program to run for the above mentioned devices.

When the power of device is off, the data in the DRAM bank is removed, that is why DRAM is name as volatile memory. Figure 1-1 shows a cell of DRAM is with the structure of a transistor connected to a capacitor or 1T1C structure.

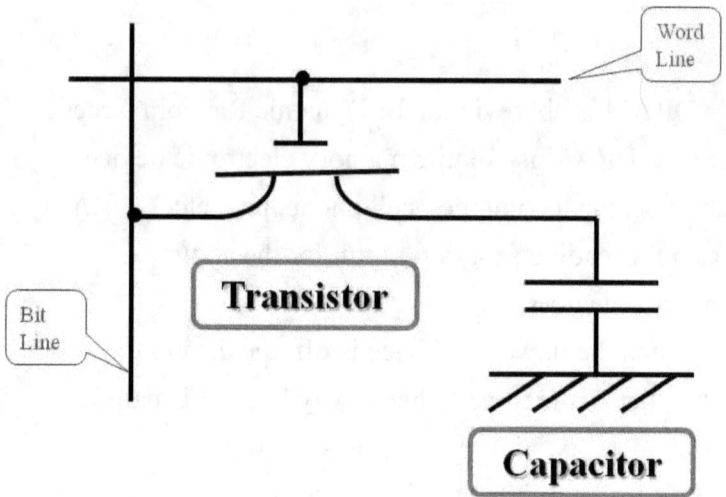

Figure 1-1. A cell of DRAM is with the structure of a transistor connected to a capacitor.

1-2. The physical limit for DRAM

The physical limit for DRAM is somewhere around 10nm. Today's most advanced devices are based on roughly 18nm to 15nm processes.

The limitation is caused by the reason that it's becoming more difficult to scale or shrink the capacitor at each node.

1-3. Cross Section of DRAM

Figure 1-2 shows a DRAM cross section view graph.

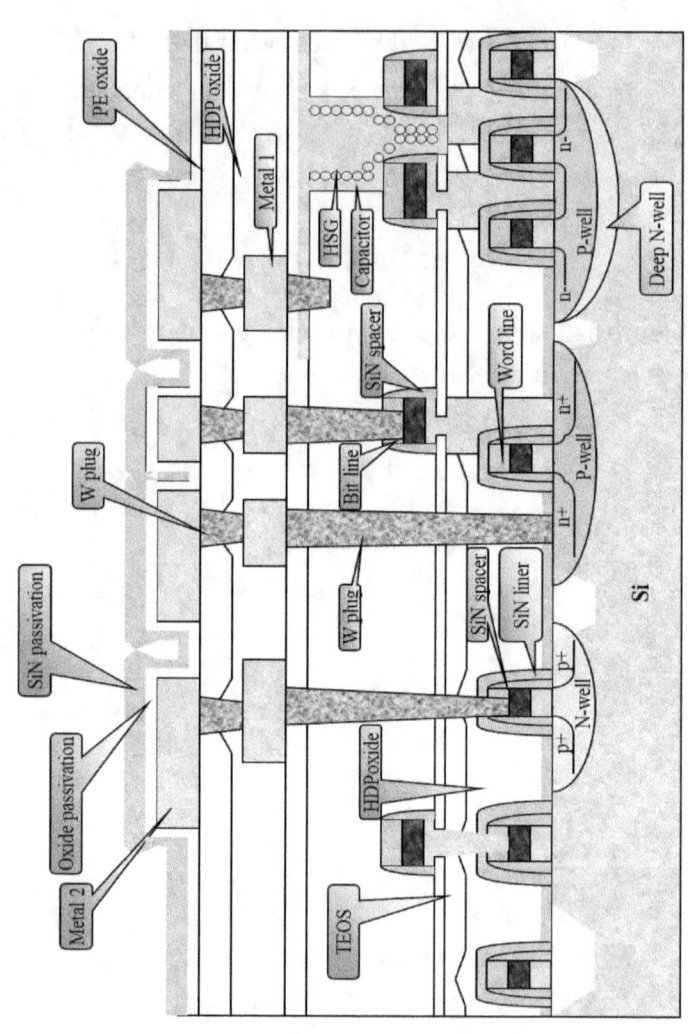

PE oxide

HDP oxide

Metal 1

HSG

Capacitor

SiN spacer

Bit line

Word line

n−

P-well

Deep N-well

n−

n+

P-well

n+

W plug

W plug

SiN spacer

SiN liner

p+

N-well

p+

SiN passivation

Oxide passivation

Metal 2

HDP oxide

TEOS

Si

Figure 1-2. A DRAM cross section view graph.

The process is from bottom Si to the top and DRAM process is a few hundred steps, however, those steps can be classified to a few key modules as follows.

(1) Isolation process: LOCOS (local oxidation of silicon) or STI (shallow trench isolation).

(2) Active area (AA): ion implantation process is proceeded after isolation.

(3) Transistor gate: word line (P1) using polysilicon (Poly) and tungsten silicide (WSi).

(4) Contact for metal line: contact hole (C1) process to soucre and drain of transistor.

(5) Bit line: the second poly line (P2) using polysilicon (Poly) and tungsten silicide (WSi).

(6) Contact for metal line: contact hole (C2) process to bit line.

(7) Capacitor process: two polysilicon layers (P3 and P4) are processed with the capacitor structure between P3 and

P4 layers. HSG process: to increase the capacitor using HSG (hemi-sphere grain) process.

(8) ONO process: three layers of silicon oxide, silicon nitride and silicon oxide for the capacitor.

(9) Metal line porces: metal 1 (M1) and metal 2 (M2) are processed then via 1 (V1) is between M1 and M2 for connection.

(10) Passivation process: final process of silicon nitride for IC protection.

The detailed description and explanation using top view and cross section view graph is in the book of the author: "DRAM-Dynamic Random Access Memory--The memory of computer, smart phone and notebook PC, ISBN 9781729479346, 2018."

Chapter 2. Novel devised DRAM

with 2T0C cell structure

This chapter is to brief depict the novel cell structure
of DRAM with 2T0C cell structure.

2-1. DRAM with 2T0C cell structure

Last year, at the conference of 2020 International Electron Devices Meeting (IEDM), IMEC presented a paper on a novel capacitor-less DRAM cell architecture, which is DRAM cell with 2T0C cell structure.

2-1. DRAM with 2T0C cell structure

The need for a storage capacitor will limit the high-density DRAM shrinking in size.

IMEC has devised a DRAM cell architecture that implements two indium-gallium-zinc-oxide thin-film transistors (IGZO-TFTs) and no storage capacitor. DRAM cells in a 2T0C (2 transistor 0 capacitor) configuration show a retention time longer than 400s for different cell dimensions. This in turn reduces the memory's refresh rate and power consumption.

The ability to process IGZO-TFTs in the back-end-of-line (BEOL) manufacturing line reduces the cell's footprint and opens the possibility of stacking individual cells.

References

1. Capacitor-less DRAM,
https://semiengineering.com/manufacturing-bits-feb-2-2/.

"Besides the long retention time, IGZO-TFT-based DRAM cells present a second major advantage over current DRAM technologies. Unlike Si, IGZO-TFT transistors can be fabricated at relatively low temperatures and are thus compatible with BEOL processing.

This allows us to move the periphery of the DRAM memory cell under the memory array, which significantly reduces the footprint of the memory die. In addition, the BEOL processing opens routes towards stacking individual DRAM cells, hence enabling 3D-DRAM architectures.

Our breakthrough solution will help tearing down the so-called memory wall, allowing DRAM memories to continue playing a crucial role in demanding applications such as cloud computing and artificial intelligence," said Gouri Sankar Kar, program director at IMEC.

Index

A

B

C

O

ONO, 10

P

P1, 10

P2, 10

P3, 10

P4, 10

Poly, 10

S

STI, 10

T

TFT, 12

Copyright

Cross Section of

DRAM
(Dynamic Random
Access Memory)
Kung Linliu, Ph. D.

Disclaimer

More professional books from Dr. Kung Linliu

Professional: Semiconductor/IC, inkjet, printing/electronics, MEMS, RFID, FPCB, micro-LED, etc.

1. 噴墨列印技術應用−標籤標示(Label, marking and inkjet printing technology), 243pp, ISBN 1230002349671, 2018.

2. 噴墨列印技術(Inkjet printing technology), 326 頁, ISBN1230002367361, 2018.

3. Inkjet printing technology for manufacturing the flexible printed circuit board (FPCB), 351pp, ISBN9781729228432, 2018.

4. 簡易影片製作方法&免費軟體-Using Windows Movie Maker-142 頁, ISBN 1230002871639, 2018.

5. Print on demand (POD) with a continuous inkjet printer, 211pp, ISBN9781729231203, 2018.

6. VLSI 製程技術(VLSI Technology), 510pp, ISBN1230002362052, 2018.

7. 半導體製程:微影與蝕刻(Semiconductor process:

Lithography and etching), 391pp, ISBN
1230002361451, 2018.

8. A small business of print on demand, 163pp, ISBN
   1230002694429, 2018.

9. Inkjet printing technology, 144 頁, ISBN
   9781729189818, 2018.

10. Successful business plan for raising 8 million US
    dollars, 95pp, ISBN1230002711935, 2018.

11. Make it simple! The secret of raising a startup fund.
    Business plan-Million USD, 76pp, ISBN
    9781729184073, 2018.

12. A perfect display! CRT, LCD, OLED and Micro LED,
    111pp, ISBN 9781729315774, 2018.

13. DRAM-Dynamic Random Access Memory--The
    memory of computer, smart phone and notebook
    PC, 184 頁, ISBN 9781729479346, 2018.

14. RFID applications and manufacturing process,
    216pp, ISBN 9781731128027, 2018.

15. Planting organic coffee tree at balcony-more
    happiness propagating life, 128 頁, ISBN

9781730892042, 2018.

16. 微光譜儀與微光柵(Micro-spectrometer and micro-grating), 102pp, ISBN 1230002889405, 2018.

17. Micro-LED Display, 140pp, ISBN 9781790838462, 2018.

18. 半導體:蝕刻製程(Semiconductor: Etching Process), 248pp, ISBN1230003042175, 2019.

19. Foldable screen smart phone using touch sensor, ISBN 9781795094900, 143pp, 2019.

20. Inkjet printing cartridge nozzle plate-Invention and lesson learned- (1), 101pp, ISBN 9781072098072, 2019.

21. Artificial gravity, centrifugal force & wafer planarization-Invention & lesson learned- (2) ISBN 9798613758050, 95pp, 2019.

22. DRAM (Dynamic Random Access Memory) process flow, 55pp, ISBN 9798610875019, 2020.

23. Semiconductor IC plasma dry etching process, 57pp, ISBN 9798612696827, 2020/2/20.

24. Micro-LED Display Process, 60pp, ISBN

9798613229710, 2020.

25. Basic Ink Jet Printing, 79pp, ISBN 9798613677917, 2020.

26. Ink Jet Printing IC Process, 92pp, ISBN 9798614525262, 2020.

27. Nano silver ink jet process for printed electronics, 40pp, ISBN 9798615855559, 2020.

28. Micro-Grating Design and Process, 72pp, ISBN 9798617520264, 2020.

29. Color ink jet disk (IJD) for mass data storage, 74pp, ISBN 9798620364121, 2020.

30. Project IJD: High Density Data Storage, 63pp, ISBN 9798622571244, 2020.

31. 微光譜儀與新冠肺炎, 67 頁, ISBN: 1230003902318, 2020.

32. Micro-Spectrometer and New Coronavirus Disease 2019 (COVID-19), 67pp, ISBN 9798646459177, 2020.

33. Semiconductor IC process and anti-fake market, 62pp, ISBN: 9798651795208, 2020.

34. A simple LED lighting circuit with Nano silver ink pen, 61pp, ISBN: 9798651005499, 2020.

35. 半導體 IC 產業與設備投資(Semiconductor IC Needs more Money Always), 163 頁, ASIN: B08B489XQX, 2020.

36. 半導體 IC 微影與全像黃光製程(Semiconductor IC lithographic and holographic patterning process), 100 頁, ASIN: B08CH4FTNF, 2020.

37. Ink Jet Printing FPCB Heater Process, 144pp, ASIN: B08CS4BKZZ/ISBN 9798665854755, 2020.

38. Ink Jet Printing RFID (Radio Frequency Identification) Process, 137pp, ISBN 9798666074039, 2020.

39. 鎳與銀金屬缺陷的消除魔法, 115 頁, ASIN:B08DZY82Q3, 2020.

40. The magic elimination of metal defects by nano-silver particle on nickel plate, 120pp, ISBN 9798671580761, 2020.

41. An economical automated R2R Flexible PCB manufacturing process, 111pp, ISBN

9798672077888, 2020.

42. Foundry, IDM and Fabless IC design house Competitive Analysis--AMD's luck over Intel is up to TSMC! 60pp, ISBN 9798673436875, 2020.

43. 半導體蝕刻全像圖形法(Plasma dry etching of hologram TSL pattern) 114 頁, ASIN: B08FRW92KC, 2020.

44. Semiconductor IC Plasma dry etching hologram pattern, 132pp, ISBN: 9798677346088, 2020.

45. Hologram seamless patterning, searching and identification of yield killer, 113pp, ISBN: 9798550306710, 2020.

46. IC high yield, low yield and yield killer, 114pp, ISBN 9798552370795, 2020.

47. Design a ring HDP semiconductor Plasma dry etcher, 132pp, ISBN 9798560791438, 2020.

48. Design a cylinder HDP semiconductor Plasma dry etcher, 134pp, ISBN 9798561001697, 2020.

49. Design a slit HDP semiconductor Plasma dry etcher, 138pp, ISBN 9798561298431, 2020.

50. Polymer blends critical phenomena for lithographic patterning, 163pp, ISBN 9798576299416, 2020.

51. Centrifugal gravity for polymer characteristics and phase separation, 163pp, ISBN 9798578148255, 2020.

52. 半導體製程技術(Semiconductor Process Technology)/中文第一版, 新版-VLSI 製程技術, 552 頁, ASIN: B08Q3D6S7X, 2020.

53. 噴墨列印技術標籤標示應用(Label, Marking and Inkjet Print Technology)/中文第一版，新版-噴墨列印應用標籤標示, 295 頁, ASIN: B08Q7GN7D4, 2020.

54. 微光譜儀與醫用檢測/中文第一版,新版-微光譜儀與新冠肺炎, 101 頁, ASIN: B08QDZ1M3H, 2020.

55. 噴墨列印技術與應用(Ink Jet Printing Technology and Applications)/中文第一版，新版-噴墨列印技術, 343 頁, ASIN:B08QDXRJ22, 2020.

56. 微光譜儀技術與應用/中文第一版,新版-微光譜儀與微光柵(Micro-spectrometer and micro-grating), 122 頁, ASIN: B08QDZSMTT, 2020

57. 半導體蝕刻製程與應用/中文第一版，新版-半導體:蝕刻製程(Semiconductor: Etching Process), 292 頁, ASIN: B08QHCNWVL, 2020.

58. 半導體微影與蝕刻製程(Semiconductor process: Lithography and etching)/中文第一版，新版-半導體製程:微影與蝕刻, 441 頁, ASIN: B08QJNC8FH, 2020.

59. Movie Maker 與影片製作方法/中文第一版,新版-簡易影片製作方法&免費軟體-Using Windows Movie Maker-161 頁, ASIN: B08QGK1P6F, 2020/12/14.

60. Cross Section of DRAM (Dynamic Random Access Memory), 23pp, ASIN: B08RYL5GHD, 2021/1/4.

61. 向植物學習-生命意義/中文第一版,新版-向植物學習-生之道(Lessons learned from plants), 140 頁, ASIN: B08QGCM133, 2020.

62. 半導體環狀高密度電漿蝕刻機臺/Design a ring HDP semiconductor Plasma dry etcher, 125 頁, ASIN: B08T8MJHPK, 2021/1/17.

63. 簡易-半導體 IC 製程, 112 頁, ASIN: B08TB3ZP31, 2021/1/18.

64. 超簡單–半導體 IC 製程, 78 頁, ASIN: B08TB4K86B, 2021/1/18.
65. 入門–半導體 IC 製程, 61 頁, ASIN:, 2021/1/18.
66. 半導體圓筒狀高密度電漿蝕刻機臺/Design a cylinder HDP semiconductor Plasma dry etcher, 126pp, ASIN: B08VH435TN, 2021/1/31.
67. 半導體縫狀高密度電漿蝕刻機臺/Design a slit HDP semiconductor Plasma dry etcher, 135pp, ASIN: B08VHR2GNJ, 2021/2/1.
68. DRAM 剖視圖/Cross section of DRAM (Dynamic Random Access Memory), 2*pp, ASIN:, 2021/2/?4.
69. Create a good business and become rich, 6*pp, ISBN *, 2021.
70. Metal roller patterning and more electroforming copies, 6*pp, ISBN *, 2021.
71. True seamless patterning and metal engraving, 6*pp, ISBN *, 2021.
72. Neat seamless patterning and more profits, 6*pp, ISBN *, 2021.
73. Seamless patterning and embossing printing, 6*pp,

ISBN *, 2021.

74. Seamless patterning and embossing-World first technology (1), 6*pp, ISBN *, 2021.

75. True seamless patterning and metal engraving-World first technology (2), 6*pp, ISBN *, 2021.

76. Writing books cannot become rich unless you are famous or a magician, 6*pp, ISBN *, 2021.

77. Solvent inkjet ink for wide format output, 6*pp, ISBN *, 2021.

78. Chemistry of Semiconductor Process, 6*pp, ISBN *, 2021.

www.ingramcontent.com/pod-product-compliance
Lightning Source LLC
Chambersburg PA
CBHW071123220526
45467CB00004B/2029